# THE EVERYDAY MATHS

# LIAT BERDUGO

Anomalous Press
www.anomalouspress.org

# Acknowledgements

I am honored to have been chosen as part of the new Anomalous Press chapbook series, and am especially indebted to Erica Mena and the Anomalous Press staff, designers, and editors for bringing this volume to completion.

My immeasurable thanks for the feedback and encouragement I have received from friends, family, and fellow artists over the past years. Thank you to Jeff Hoffstein and the National Science Foundation, who jointly gave me the opportunity to spend my days editing mathematics textbooks. My very special thanks also goes to Thalia Field, whose support and feedback has shaped this manuscript from its inception and who always asked that I flex my finishing muscle.

*The Everyday Maths*
© 2013 by Liat Berdugo

First edition 2013. Printed in the USA.
ISBN: 978-1-939781-03-1

Designed and typeset by Sarah Seldomridge.

This is available as an ebook & audiobook from Anomalous Press.
**www.anomalouspress.org/books/maths.php**

# CONTENTS

| | | | |
|---|---|---|---|
| JUST THE HEADS. | 1 | ALL MEN EQUAL. | 25 |
| FOR THE SAKE OF TRANSPARENCY. | 2 | BIOPSY. | 26 |
| DEXTRAL DOMINANCE. | 3 | PATH OF REFUSE. | 27 |
| CROP CIRCLES. | 4 | COOKIE SHEET. | 28 |
| TO PASS THE TIME. | 5 | SOLDIERS FOR A CAUSE. | 29 |
| IT COULD BE BETTER. | 6 | COMPLEXITY SHMOMPLEXITY. | 30 |
| BARBIE DOLL. | 7 | BANG. | 31 |
| SOMETHING DIFFERENT PLEASE. | 8 | SKIN COLOR. | 32 |
| COLOR BY NUMBER. | 9 | INFORMATION COMMERCIAL. | 33 |
| BETTER. | 10 | FIDELITY INVESTMENT. | 34 |
| A VISUAL CORTEX, DAMAGED. | 11 | AGEING. | 35 |
| PLEASE ORGANIZE. | 12 | CONVERTIBLE. | 36 |
| A SUPERSTITION. | 13 | SHOPPING SPREE. | 37 |
| WHEN SOME THINGS FAIL. | 14 | PLEASE HOLD HANDS. | 38 |
| DECISIONS, DECISIONS. | 15 | SOFT PRETZEL. | 39 |
| HORSEBACK. | 16 | A BURGEONING APPETITE. | 40 |
| RELIABLE WITNESS. | 17 | ROSWELL, NEW MEXICO. | 41 |
| A COMMAND. | 18 | OPPOSITES. | 42 |
| CAGED BIRD. | 19 | ANOTHER GO. | 43 |
| AT FIRST SIGHT. | 20 | THERE MIGHT HAVE BEEN SOME MASTICATION. | 44 |
| THAT THING AHEAD. | 21 | SENSE OF DIRECTION. | 45 |
| SUNNY SIDE UP. | 22 | PICK-UP STICKS. | 46 |
| FLOWER ARRANGEMENT. | 23 | BEDTIME ARRANGEMENTS. | 47 |
| DOUGHNUT. | 24 | | |

**FOR CLAIRE**

# JUST THE HEADS.

They are watching a movie. They are posing for civil union pictures. Apart, they are nothing. Together, they are everything.

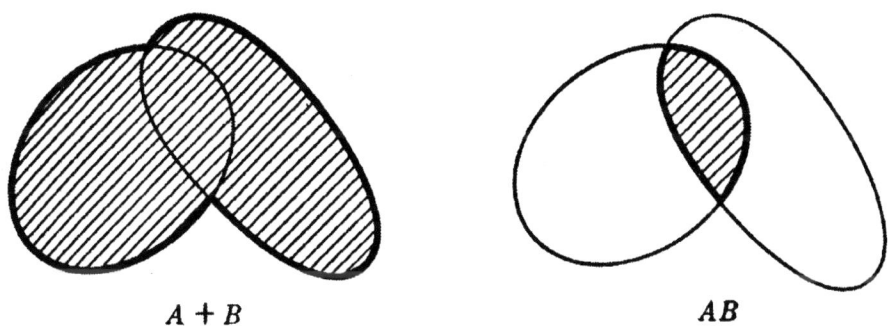

Figure 79. Everything and a part of everything.

# FOR THE SAKE OF TRANSPARENCY.

They trained the youth to look through. To read between, to see beyond. It was a difficult skill to master.

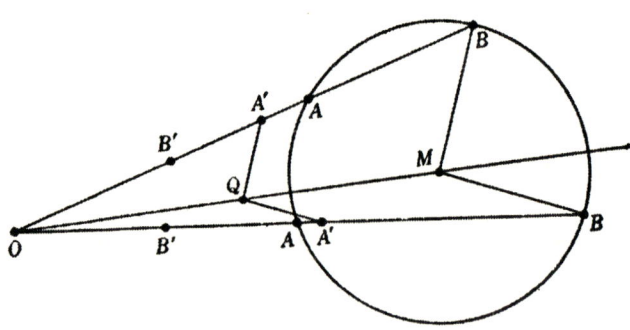

Figure 110. The cleverest could see the moon behind the airplane.

# DEXTRAL DOMINANCE.

One day, seemingly out of nowhere, he became right-handed. It took him some time to reorient.

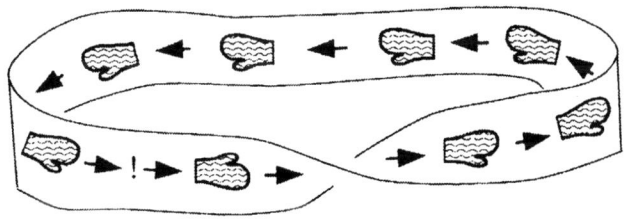

Figure 148. Exclamation mark indeed.

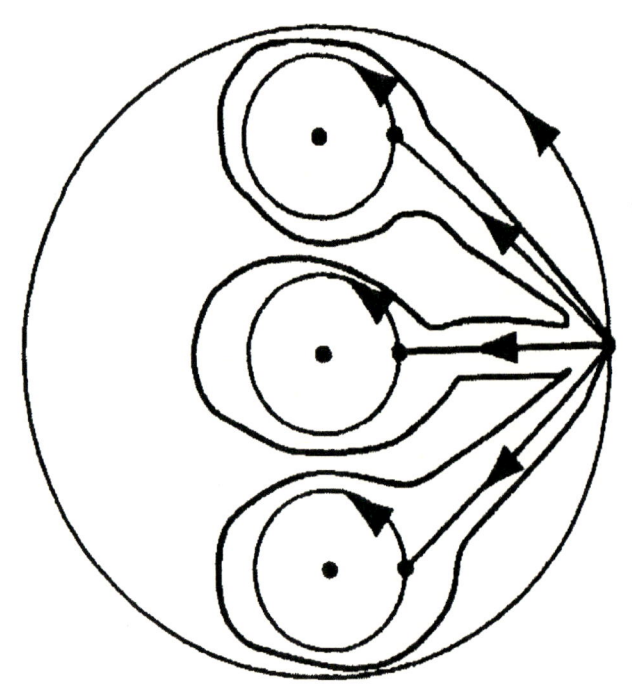

# CROP CIRCLES.

He awoke to find a design carved into his corn field. I am not new-agey, he said, but who else could have done this. They do not have the steadiest hands, though.

Figure 108. Santa Rosa, CA.

# TO PASS THE TIME.

She would wait. She would become bored. At times like these, she would think of metronome patterns inescapably.

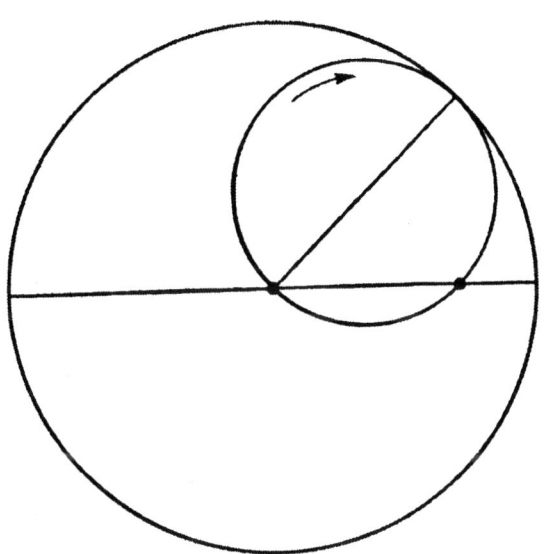

Figure 63. Circular thinking.

# IT COULD BE BETTER.

They wanted more cusps. Mr. Cartographer, they said, please add some kinks to the boundary of our country. We would like some things to be complicated.

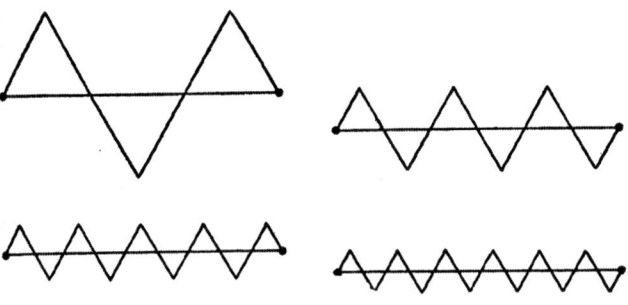

Figure 49. Prospective borders.

# BARBIE DOLL.

Don't be fooled, she said. No one looks like that. Bodies come in all shapes and sizes.

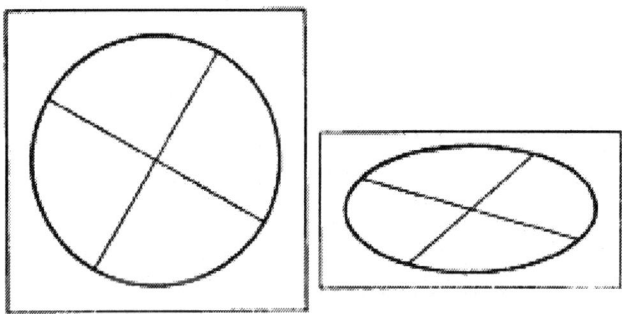

Figure 97. Also, do not tumble dry.

7

# SOMETHING DIFFERENT PLEASE.

At a time when he craved heterogeneity, looking towards his friends was less than comforting.

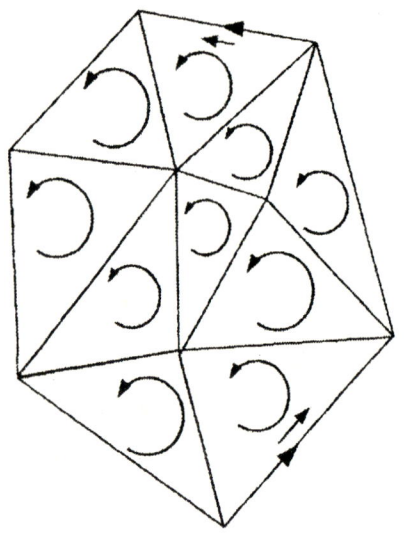

Figure 12. As if counterclockwise were the new counter-revolution.

# COLOR BY NUMBER.

What it lacked in variation it made up in its introduction of free will.

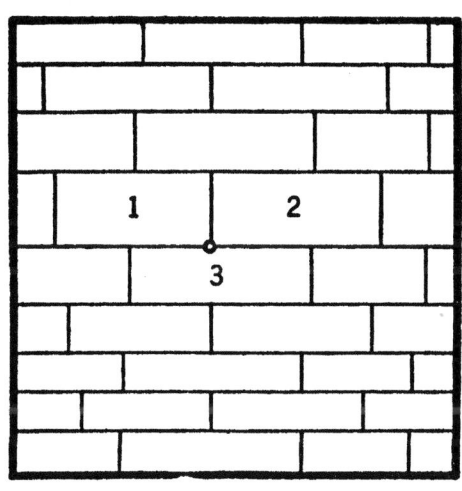

Figure 19. 1: Maroon. 2: Maroon. 3: Burnt sienna. Unlabeled: Your choice of maroon or burnt sienna.

# BETTER.

Yes, we like these much more. Our foes will not know what is inside and what is outside. We shall enact them immediately.

Figure 49 (ii).
Prospective borders.

# A VISUAL CORTEX, DAMAGED.

When he fixated on the faces of others, all distinguishable features disintegrated, leaving only the perimeters.

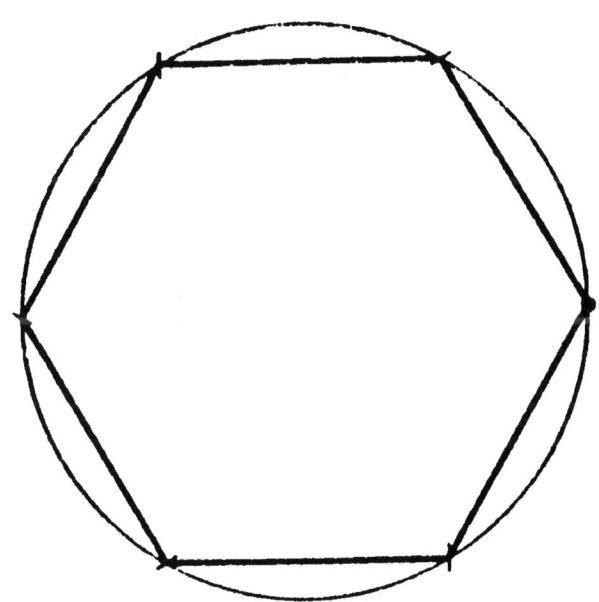

Figure 75. She was so beautiful.

# PLEASE ORGANIZE.

To promote order, they chose to untie the knots, beginning with the first knot. Their dexterous hands became tired, as one might imagine.

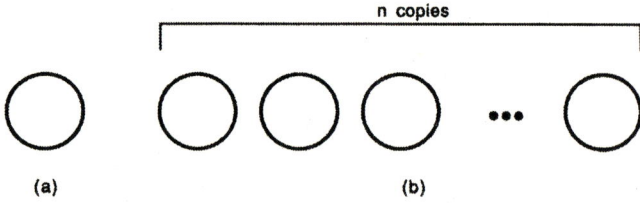

Figure 2. (a) The unknot. (b) Many unknots, with some of them omitted.

# A SUPERSTITION.

Concavity leads to cavities, said the dentist. Therefore, draw all your pictures convex, like so. You will thank me.

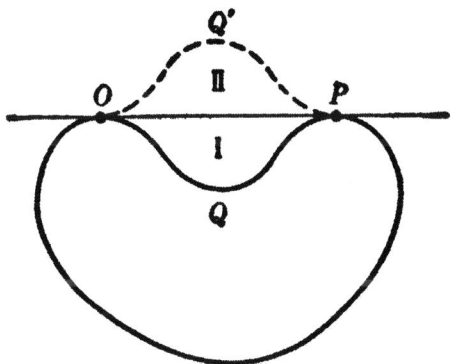

Figure 4. As necessary as flossing.

# WHEN SOME THINGS FAIL.

Someone has come and labeled everything. We are at a loss for words.

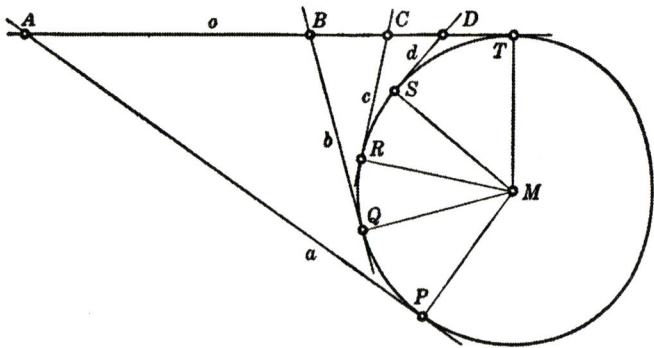

Figure 9. A word machine, albeit deficient in vowels, used via repeated connecting of the dots.

# DECISIONS, DECISIONS.

He could not decide which to wear. All fit his nose so perfectly.

  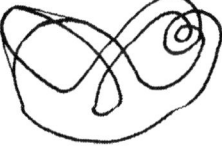

Figure 22. Masking identity.

# HORSEBACK.

When mounting, they explained, take care to place your buttocks squarely on the saddle. This is of utmost importance.

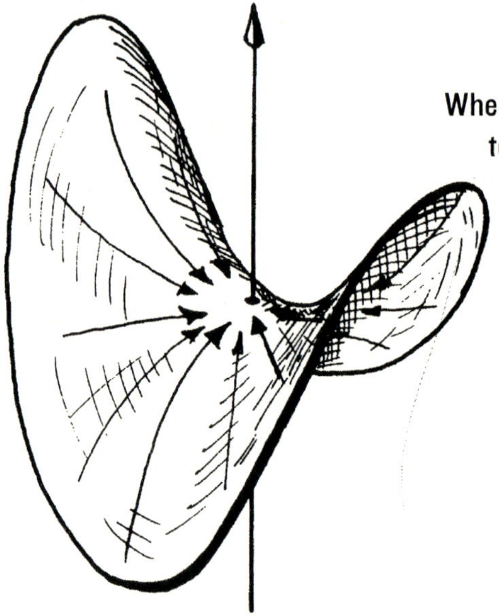

Figure 134. Even if the horse is elsewhere.

# RELIABLE WITNESS.

They asked, which way did he run, sir?
We expect a definite response.

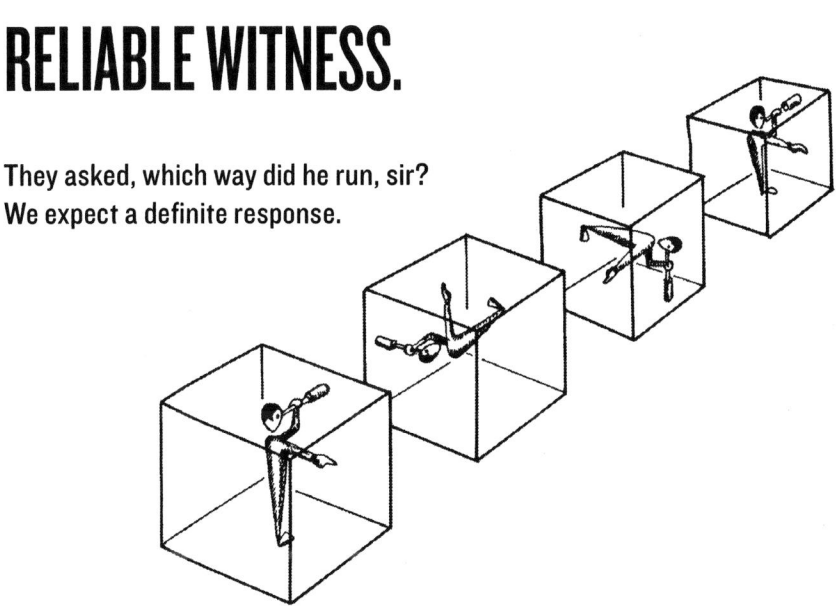

Figure 70. Evasion.

# A COMMAND.

Like a cornucopia, he said. As numerous as the sands of the earth.

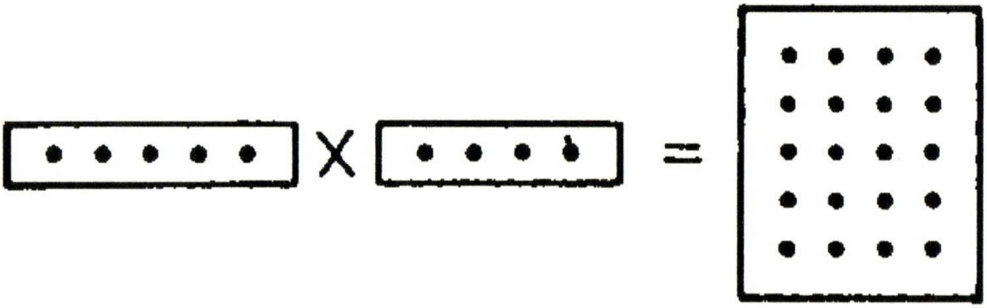

Figure II. Be fruitful.

# CAGED BIRD.

It broke his heart to see such a beautiful and long-necked creature boxed away like that.

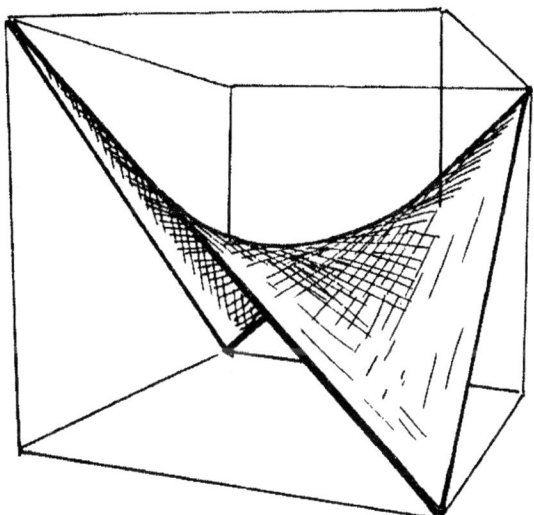

Figure 135. Zoo of the abstract.

# AT FIRST SIGHT.

They met in a way not uncommon. The rest was history.

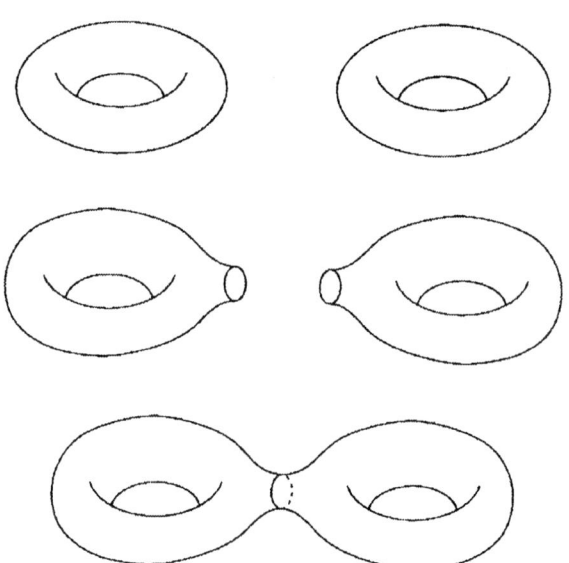

Figure 44. Smooch.

# THAT THING AHEAD.

He tried to bypass the thing ahead. But when he came out, he was only a portion of the man he used to be.

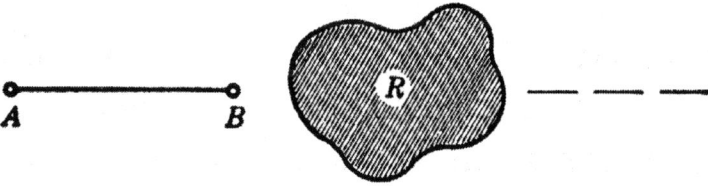

Figure 4I. Continuation beyond the specified obstacle is spotty at times.

# SUNNY SIDE UP.

He was a precise man. He preferred to carefully allow the yolk to run onto his plate of morning potatoes. It was a surgical strike.

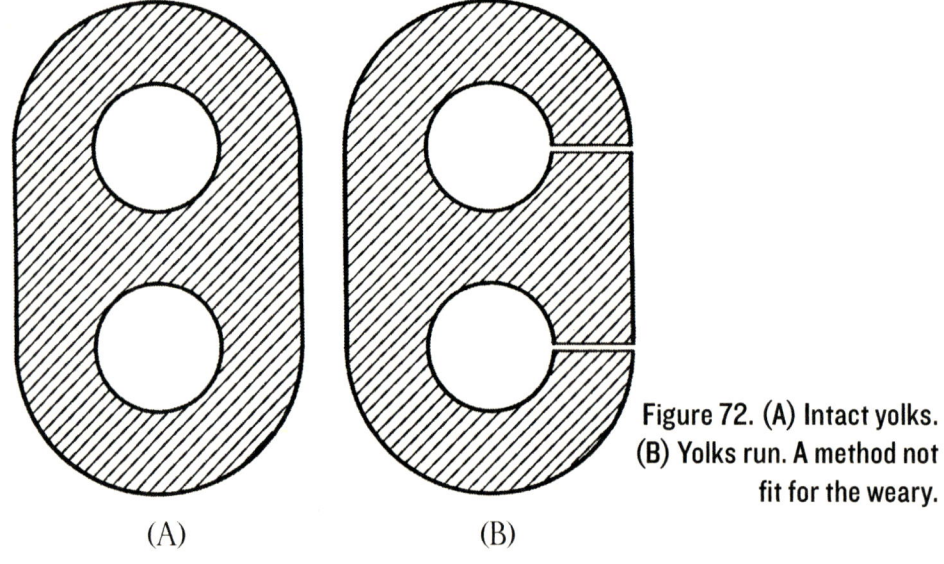

Figure 72. **(A)** Intact yolks. **(B)** Yolks run. A method not fit for the weary.

# FLOWER ARRANGEMENT.

Yes, it is a nice vase, she said. Where ought I pour the water, though?

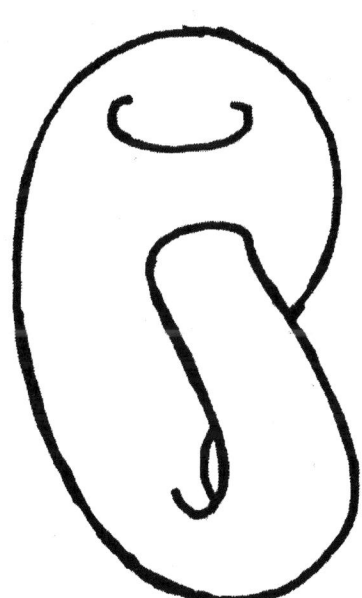

Figure 147. When insides are outside.

# DOUGHNUT.

This is attractive, she said, but I asked for one with sprinkles.

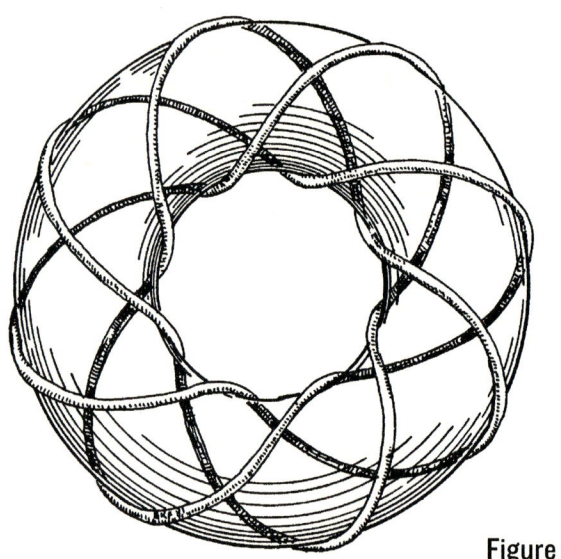

Figure 56. Customer satisfaction eluded.

# ALL MEN EQUAL.

When it came right down to it, although they each received an extra sex chromosome in utero, they were both just men. Men who liked to chop firewood and trace their roots with diagrams.

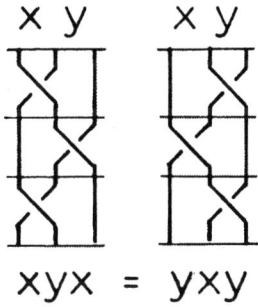

Figure 29. They held those truths to be self-evident.

# BIOPSY.

It could be melanoma, said the doctor. I would like to send a sample to the lab for scrutiny.

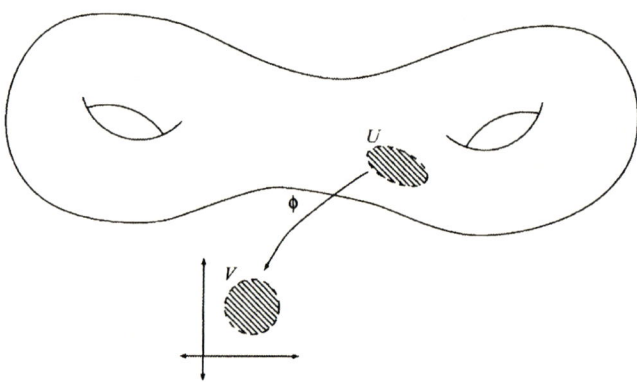

Figure 99. Under the microscope, professionals scrutinized it more roundly.

# PATH OF REFUSE.

The problem, ma'am, is that your drainage pipe became quite friendly with that of your neighbor. The situation will be difficult to straighten.

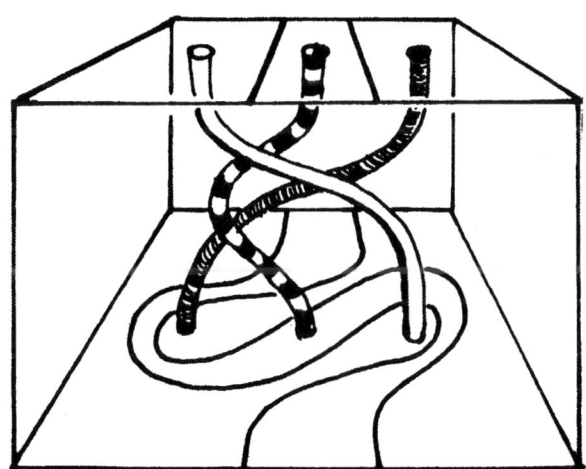

Figure 141. A plumber's prognosis.

# COOKIE SHEET.

I will place the dough wherever I choose, he said. The two-inch-apart rule is so yesterday. Also, this way it looks like mitosis.

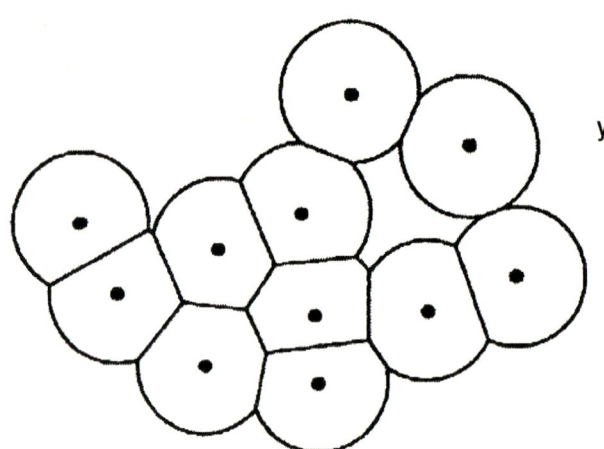

Figure 93. A renewed freedom of expression.

# SOLDIERS FOR A CAUSE.

Like many, they wished to be a part of something larger than themselves.

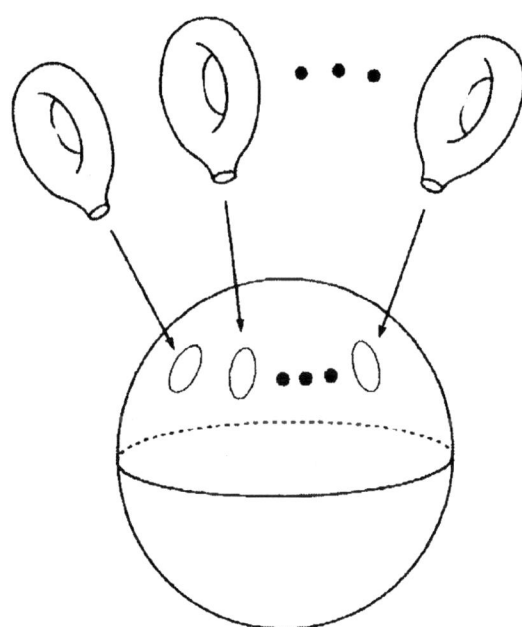

Figure 32. The world is calling.

# COMPLEXITY SHMOMPLEXITY.

He never could draw a heart. He tried many times. It is too complex,
he said, and can't you still tell what I'm going for?

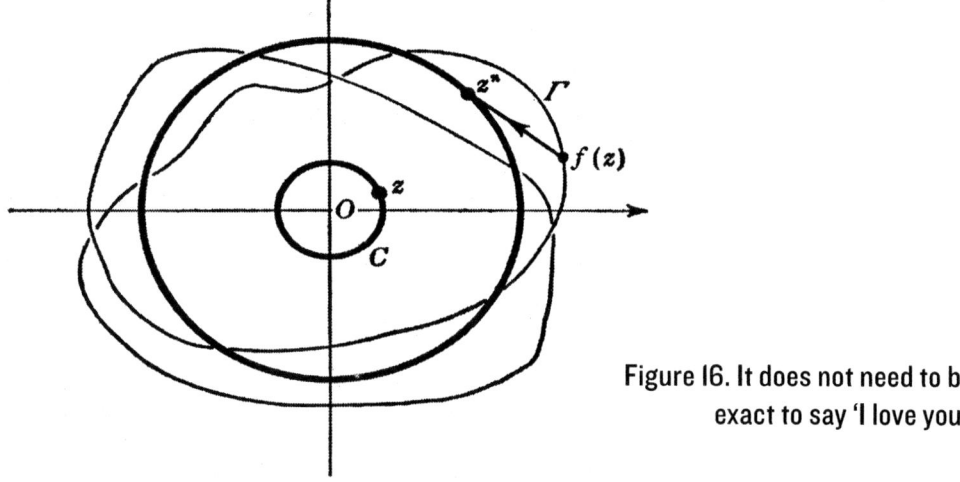

Figure 16. It does not need to be
exact to say 'I love you.'

# BANG.

We think that the universe exploded at some point, he said. We further believe that it happened in the following very specific order.

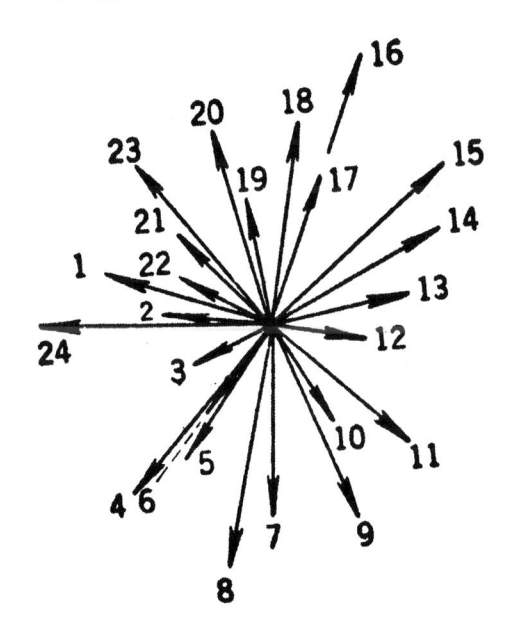

Figure 66. Just a hypothesis.

# SKIN COLOR.

She found it difficult to be in the obvious minority.

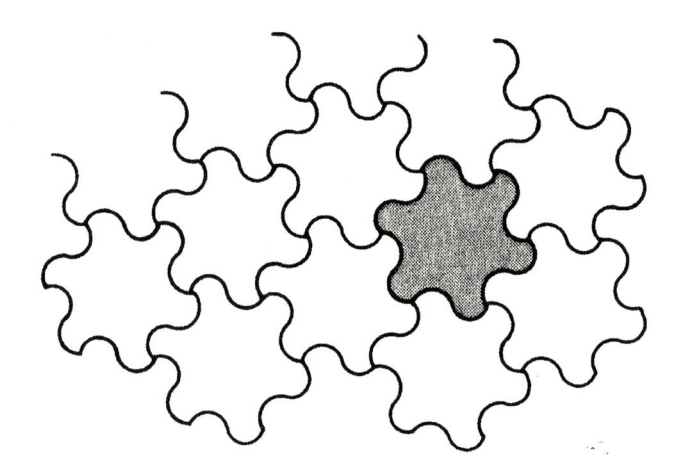

Figure 5. Fitting in.

# INFORMATION COMMERCIAL.

Lose that boxy shape! Get definition where you need it most! Our ab-revolution belt does just that. Strap it on and relax in front of your TV.

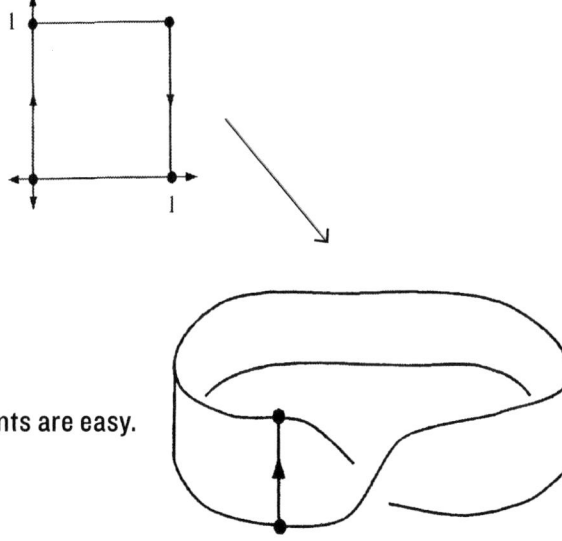

Figure I32. The three payments are easy.

# FIDELITY INVESTMENT.

These days, she explained, relationship models are more complex. We call it p o l y a m o r y. Are you interested in purchasing a share?

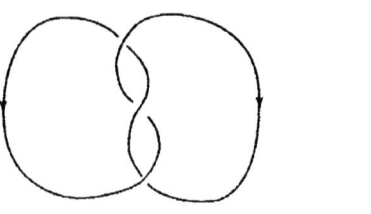

Figure 92. Stay in the loop.

# AGEING.

Most frequently, balding begins in a roughly ovular region. This is seen most clearly when we contrast the bald area with texture, which it lacks.

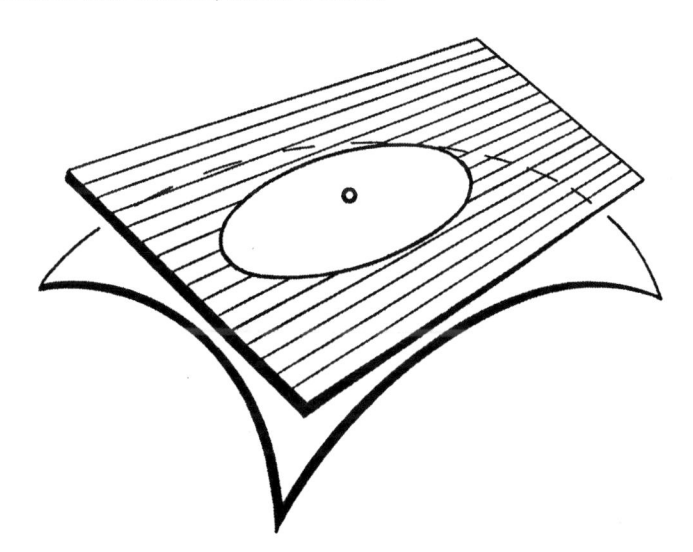

Figure 31. A soft spot.

# CONVERTIBLE.

He wanted a big change in his life, but some changes were simply unallowable.

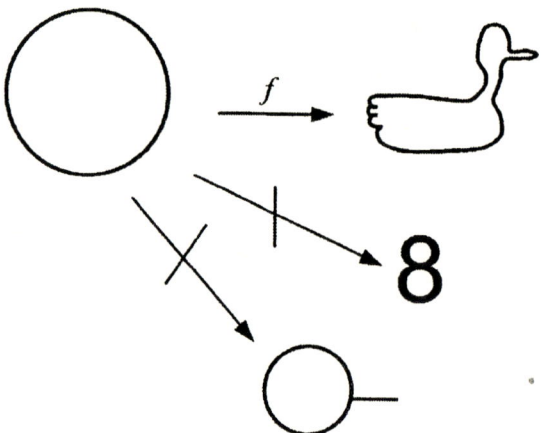

Figure 78. But he was no quacksalver.

# SHOPPING SPREE.

This is a nice bonnet indeed. What sizes does it come in, she asked.

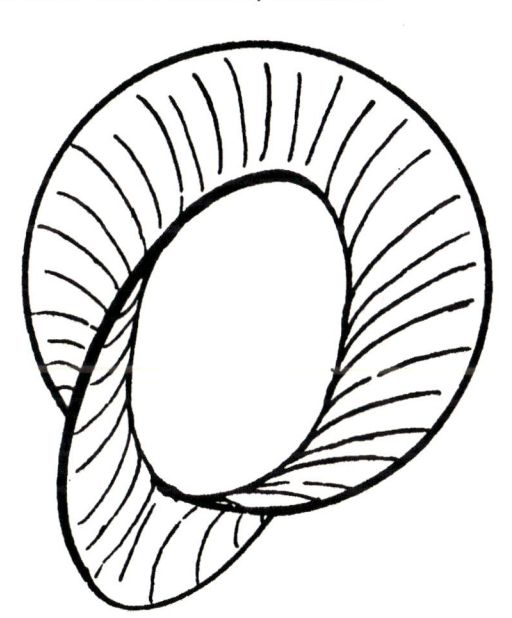

Figure 67. Modernism for the Amish.

# PLEASE HOLD HANDS.

Their therapist suggested they find intimacy manually. After their first attempt, they could not go back to the way they were previously.

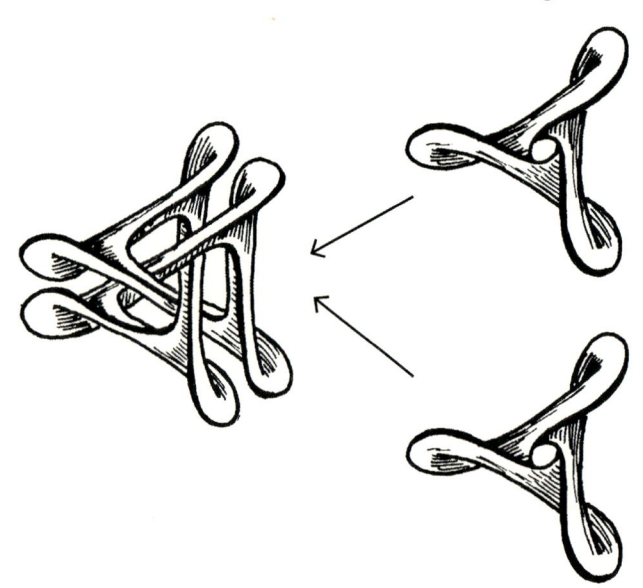

Figure 83. Independence lost.

# SOFT PRETZEL.

We have seen a marked decrease in output, she said. Our most recent production is miniscule.

Figure 102. Bring in the consultants.

# A BURGEONING APPETITE.

After a large dinner, she sometimes needed to unbutton her pants.

Figure 56. It went straight to her hips.

# ROSWELL, NEW MEXICO.

They took nothing. In fact, they left a few things behind. We have only a spotty understanding of what these things are, however.

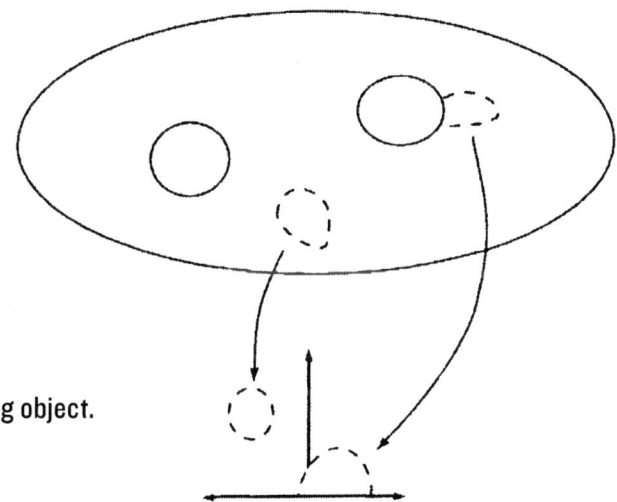

Figure 13. Unidentified flying object.

# OPPOSITES.

This is not going to work, she said. You and I have nothing in common.

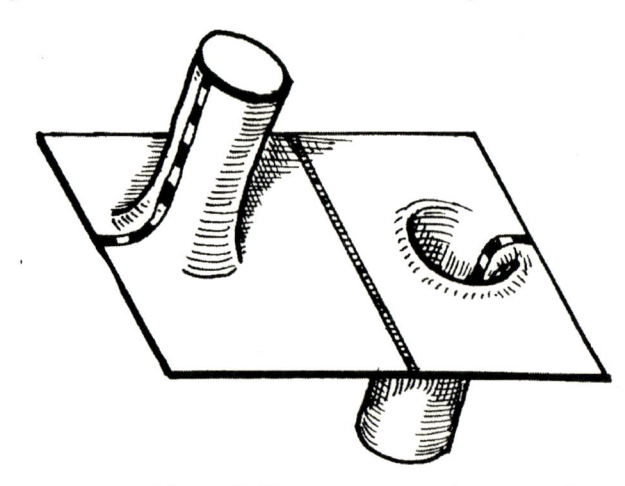

Figure 8. You say tomayto, I say tomahto.

# ANOTHER GO.

But we see eye to eye on so many issues, he said, like our opinion of draw bridges.

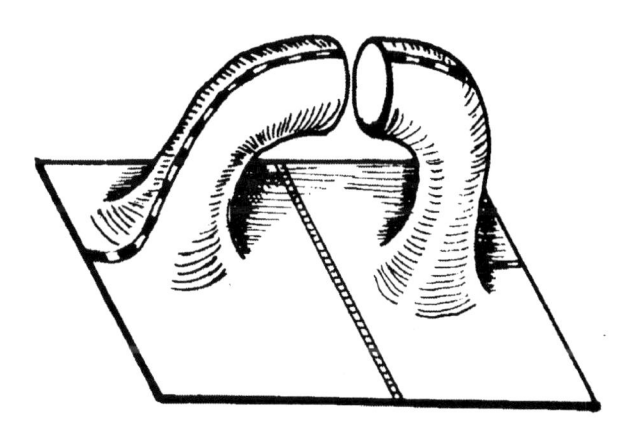

Figure III. We agree.

# THERE MIGHT HAVE BEEN SOME MASTICATION.

Perhaps the dog chewed it. Or, it melted. What is the current temperature in degrees celsius? It is difficult to determine cause and effect. At least there is still that wide opening to get your arm through.

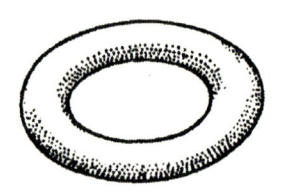

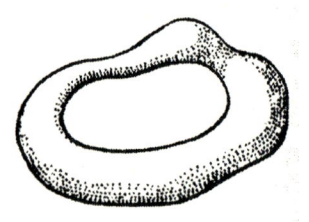

Figure 7. There is still a hole.

# SENSE OF DIRECTION.

Where do you think we might exit, he asked. I am not sure. Let us try this way.

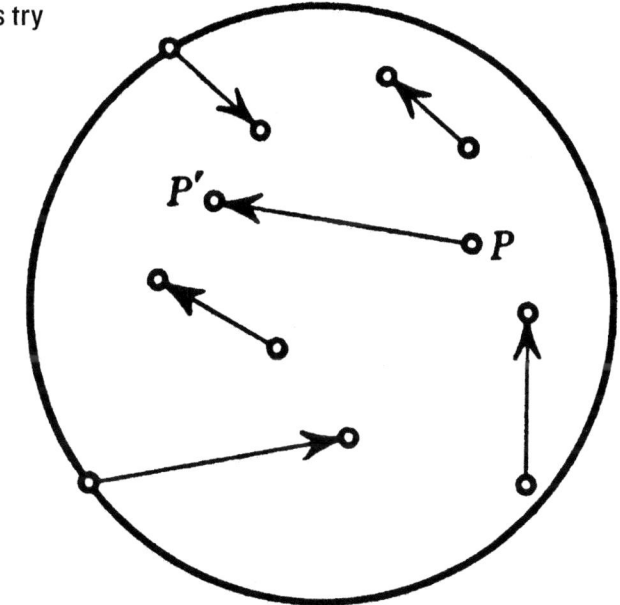

Figure 109. Petri dish.

# PICK-UP STICKS.

He sweat, not knowing which stick to remove first. In addition, the sun was shining brightly, making him sweat even more.

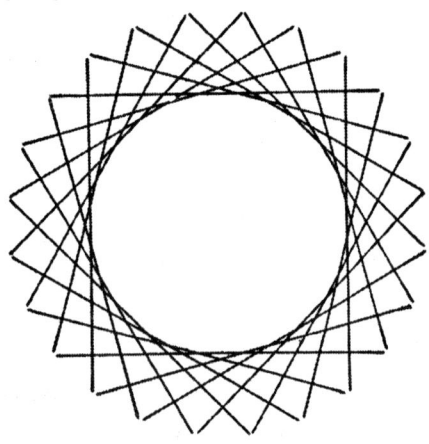

Figure 16. In between some things there is sometimes another thing.

# BEDTIME ARRANGEMENTS.

They were never good at sharing sides of the bed. Finally, they had a special bed imported from Italy which had a side for everyone. This was, in the end, the right decision.

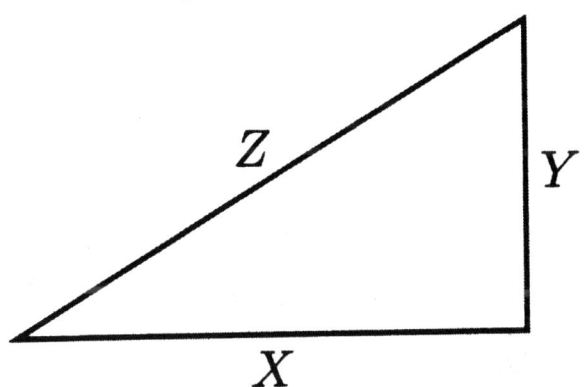

Figure 10. They slept better this way.

The winner of the first Anomalous Press chapbook contest in poetry, this manuscript was selected for publication by Cole Swensen.

The sixth book in the Anomalous Press series, this book was typeset and designed by Sarah Seldomridge in a limited edition of 200 copies.

Anomalous Press is dedicated to the diffusion of writing in the forms it can take. We're searching for imaginary solutions in this exceptional universe. We're thinking about you and that thing you wrote one time and how you showed it to us and we blushed.

www.anomalouspress.org

**Liat Berdugo** is an American artist and writer whose work focuses on the strange, delightful and increasingly ambiguous terrain between the digital and the analog, the online and the offline, and the scientific and the literary. Her work has been exhibited in galleries and festivals internationally, including the Urban Art and Media Organization in Munich, the Athens Video */ Art Festival, the Boston Cyberarts Gallery and the DysTorpia Media Series in New York. She studied mathematics at Brown University and Digital + Media at the Rhode Island School of Design. More at http://digikits.ch.

Available from **Anomalous Press**:

*An Introduction to Venantius Fortunatus for Schoolchildren or
Understanding the Medieval Concept World through Metonymy*
by Mike Schorsch

*The Continuing Adventures of Alice Spider*
by Janis Freegard

*Ghost*
by Sarah Tourjee

*The Everyday Maths* by Liat Berdugo
selected by Cole Swensen

*Mystérieuse* by Éric Suchère, translated by Sandra Doller
selected by Christian Hawkey

*Smedley's Secret Guide to World Literature
by Jonathan Levy Wainwright, IV, age 15*
by Askold Melnyczuk